I0486274

IMPRESS YOUR CAT

KNOW ABOUT MOBILE PHONES AND THEIR SAFETY

How do they work?

Can they harm me?

by

Austin Farrell

1

This is the first book
in the *'Impress Your Cat'* series

– lock out for the other titles.

Drawings of *'Tadhg'* the cat
by Ro Farrell

Published and printed by Lulu.com
Available electronically

ISBN 978-1-4466-6398-1

FOREWORD

The reason I wrote this book is because, like many others, I was concerned about the possible health impact of using mobile phones. We have all heard reports of brain tumours being linked to mobile use, and I wanted to know the facts. In looking at this question, I came across research findings which surprised me, and I wanted to share this knowledge with others.

A second reason for writing is to de-mystify the technology of mobile phones. We use the phones every day, but don't really understand what makes them tick. For many of us, technology has become an alien world, filled with jargon and mystery - we don't know how things work and think we never will. Technology is affecting our everyday lives, but we aren't quite sure how or why.

Having a background in many areas of science and engineering, I spend a lot of my time explaining technology and its implications, in an understandable way, to all sorts of people - ranging from work colleagues, to lawyers, executives and family members (and anyone unfortunate enough to ask me how something works).

I believe that technology is not just the preserve of geeks and tech-heads - we can all understand it, provided it is explained properly.

INTRODUCTION

A whole generation now takes mobile phones for granted. Talking, texting, sending photos and videos, downloading e-mails and web pages wherever we are, has become an essential part of life.

Even though we have come to depend on our phones, how they work has tended to be a bit of a mystery – and there are nagging doubts about their safety.

The mobile is a recent chapter in the long story of communication - starting with runners with clay tablets, ranging through signal fires and semaphore flags on hill-tops[1], down to the more recent telegraph and telephone.

The telegraph and telephone massively changed our society and the mobile is doing the same. Mobiles have transformed business and have made the mobile office a reality. They have evolved into a universal portable device with video, Internet, radio, TV, music - even navigation.

The mobile market is also a battleground between major players in North America, Europe and Asia. The vast market has turned unknown companies in faraway places into household names – before mobiles, who had ever heard of Nokia?

---

[1] Napoleon used a semaphore system (signalling with flags) in France, with towers 12 to 25 kilometres apart. The system - mentioned in the novel 'The Count of Monte Cristo' - was invented by **Claude Chappe** in 1792. It was superseded by the telegraph in 1846.

Nokia of Finland is the world's largest mobile maker. A wood-pulp mill, which started in 1865, later moved to the town of Nokia[2]. After World War I, the Finnish Rubber Works bought the pulp mill and a cable works and began using the Nokia brand. The cable works spawned an electronics division, which later began making two-way radios - which in turn led to mobile phones in 1987.

Some medical researchers have raised doubts about the safety of mobile phones, especially for users, who are making calls for several hours each day over many years. Of particular concern to these researchers, is the generation now growing up who may be heavy mobile users from an early age. This concern has prompted the Australian Government to fund an investigation into the possible dangers.

Given the how widespread and useful mobiles are and the doubts about their safety, we need to understand something about the technology and what it might do to us.

---

[2] The town of Nokia is called after the River Nokianvirta which runs through it. Nokia is an old Finnish word for a sable (a small animal valuable for its fur).

The **purpose** of this book is to help:

- take away some of the mystery surrounding a key technology driving social and business change; and

- be aware of some of the health and safety issues involved in using mobiles and make informed choices for ourselves and our children.

For your convenience, the book is divided into **four** parts:

- the first part is a basic outline of a how a typical mobile phone system works;

- Part 2 is a discussion of the health and safety aspects;

- the third part is a quick look at some future mobile trends; and

- Part 4 gives you more detail on how mobile phone systems operate (this is the bit that really impresses cats!)

PART 1 – THE OUTLINE

What are Mobile Phones?

A mobile phone is a **two-way radio** linked to the normal telephone system. The system is arranged so that wherever you go in a city or town or along a main road, you are mostly in mobile range. The phone you hold in your hand is supported by a largely invisible infrastructure, which costs hundreds of millions of dollars.

How did Mobiles Start?

In the beginning..... the first mobile phones were so large they could only be carried in cars. The next generation were brief-case sized units. Advances in electronics have now made them so small that they will fit in tight jeans.

The first call from a handheld mobile was made
on 3-APR-1973 in New York City
by **Martin Cooper** of Motorola.

The first 'real' mobile system in Australia was set up by Telecom Australia (now Telstra) in 1987 using the U.S. AMPS[3] analogue (ie non-digital) standard

After a slow start, the Telstra system (called 'Mobilenet'[4]) became very popular and even though the handsets were like bricks, it eventually served over a million customers. Although built and operated by Telecom Australia, the system was shared by both Optus and Telecom customers.

When the Australian Government partially deregulated the telecommunications industry in 1992, they licensed **three** mobile carriers. The licence stipulated that the AMPS system would be phased out by 2000, and all three carriers would use the European GSM[5] digital mobile system, which was the most advanced system at that time.

Each of the three original carriers (Telstra, Optus and Vodafone) built their own digital networks, which are continually evolving and expanding, to cover most of the country.

---

[3] AMPS stands for **Advanced Mobile Phone System**. It was developed by Bell Labs and released in 1983. This was a 'First Generation' (1G) system. This is not to be confused with the measurement of current - Amps.

[4] '**Mobilenet**' was the only analogue mobile phone system ever to cover an entire continent.

[5] **Global System for Mobiles**. Originally GSM stood for **Groupe Special Mobile**, the French term for the European project team which oversaw its development. GSM is s so-called 'Second Generation' (2G) system.

What do we mean when we say 'digital'?

The first electrical communication system was the telegraph. The telegraph revolutionised the 19th century world, with news from distant parts zipping over the wires in hours rather than weeks or even months.

Telegraph operators tapped out text messages in a simple code - 'Morse Code' - called after **Samuel Morse**, a telegraph pioneer (see the picture of Sam as a young man).

Morse Code uses combinations of a 'dot' or a 'dash' –for example, the distress message SOS (standing for "Save Our Souls") is represented by

· · · _ _ _ · · ·

In the digital electronic world, messages are also stored and transmitted in a code, but this time the code uses two **digits** - a **0** (zero) or a **1** (one). Everything – text, data, pictures, voice, maps, music, video – can be represented by strings of zeroes and ones.

There are internationally accepted standards for coding all these things – all we have to know is which standard is being used.

Many other mobile players have now joined in, either to build their own networks or re-sell time on the existing systems.

This description focuses on the basic GSM system, which is the world's most popular mobile phone standard. It is estimated that **3.3 billion people**[6] throughout the world use a mobile phone.

The Mobile Phone

The mobile you carry in your hand can be seen as having **three** important parts:

- the **handset** , with its keypad and display;

- a small removable chip card inside - about the size of a postage stamp - called a **SIM**[7] **card;** and

- a rechargeable **battery**.

---

[6] The number of mobiles in use at the end of 2007, according to a report by the International Telecommunications Union (ITU), issued in May 2008.

[7] Subscriber Identity Module.

The handset

The handset is a **plastic case** with the processing electronics inside, a **keypad** to dial the numbers (and input text) and a **liquid crystal display** (LCD) on the front.

Liquid Crystal Displays

Crystals are normally solid, but a type of organic material called **nematic** crystals, are liquid. When an electrical voltage is applied to a liquid crystal, it changes the way polarised light passes through it.

In a black-and-white LCD, a very thin layer of the liquid crystal is sandwiched between a reflective surface (on the bottom) and a sheet of polarising glass with transparent electrodes on the surface.

When a low voltage is applied to an electrode, the area directly underneath appears dark. By carefully choosing the shapes of the electrodes we get all letters, numbers and the other shapes we want.

Colour screens for mobiles are made like flat-screen TVs. Each picture element (or 'pixel') on the back-lit screen is composed of three tiny LCD dots, with red, green and blue filters. Each colour dot works the same way as a black-and-white dot. Any colour can be made using combinations of red, green and blue.

Complicated and very fast electronics inside the phone select the correct colour for each pixel to make up the complete picture.

The simplest black-and-white LCD screens show the state of the battery. It also shows the signal strength from the nearest base station - the dialled or calling numbers - and text messages.

The more sophisticated (and expensive) colour LCD screens in all new mobiles show all this, plus colour pictures, web pages and video.

Inside the handset is a small microphone and small loud-speaker near your ear. When you speak into the microphone during a phone conversation, your voice is sampled.

It is then converted by a small, but powerful computer, into a string of numbers, compressed and temporarily stored in a memory chip.

Your stored voice is then transmitted in short radio pulses, through the antenna (which may stick out the top, but is usually folded inside the case).

These transmitted radio pulses near your head concern some users – can these pulses cause brain tumours? There is more on this topic later.

With the received signal (which is much weaker than the transmitted signal), the reverse happens.

The radio pulses with the digital information, come down the antenna and through the radio section. Here, another small processor un-compresses the voice, which you hear as a normal voice in your phone.

The pulsing of the transmission allows other phones to share your channel. Up to seven other phones can use the same channel by taking turns (so quickly that you are not aware of it).

Also, because your phone is transmitting short pulses, with big gaps in between, it uses far less battery power.

Every mobile handset has a unique identifying code (**not** the same as the phone number) - which is useful in tracing stolen phones.

The SIM card

The **Subscriber Identity Module** or SIM card is the key to the fraud resistance and voice privacy of the phones.

The SIM card has all your details stored in it. It has special data, some used to verify your identity to the system and also some to encrypt (ie scramble) your voice for privacy.

If you put your SIM card into another handset, that handset is recognised by the system as your phone for billing purposes.

You could buy a number of SIM cards to fit in your handset - perhaps one for private use and the others for your business.

The battery

A big factor in increased battery time, is the pulsing, referred to above.

Other factors are - the power consumption of the electronics in the phone has been reduced over the years, plus the newer batteries have much more capacity.

Mobile Phone Batteries

The batteries in the first mobile phones were **Nickel-Cadmium** (Ni-Cad) which suffered from 'memory' problems - they needed to be fully discharged before re-charging for best results.

The next generation mobiles used **Nickel-Metal Hydride** (NMH) batteries - the type used in the Toyota Prius hybrid cars - which have two to three times the capacity of the same size Ni-Cads.

All new mobiles now use **Lithium-Ion** (L-I) batteries, which have the highest capacity of all. These are very light, and have a very low 'self-discharge rate' (the loss of charge when not being used) of about 5% per month.

This compares with 10% per month for Nickel-Cadmium batteries, and over 30% per month for Nickel-Metal Hydride batteries).

There is more on Lithium – Ion batteries right at the end (**see Extra Bits**).

The Very First Battery

The battery was invented in 1800 by **Allesandro Volta**, a professor of physics at the Royal School in his native Como, in northern Italy (near the Swiss border).

The electrical unit, the Volt, is named after him.

His original battery was a stack of large copper and zinc discs separated by paper soaked in salty water.

Later improvements to Allesandro's battery made possible the electric telegraph – the start of the communications revolution.

The museum in Como commemorating Volta's work is on a hillside reached by a funicular railway. On a clear day, the view over Lake Como is spectacular.

If your mobile is low on battery, try:

- turning off features you're not using – like jazzy screen savers and 'Bluetooth';

- turning off the phone in poor signal areas - like underground car parks (this stops the phone straining to try and get a signal); and

- keeping the phone cool – don't leave it in the sun or on top of something hot.

For more useful battery tips and info, try www.batteryuniversity.com.

The Rest of the Mobile Network

'Cells'

Mobile phones are sometimes referred to as *'cell-phones'*, particularly in the UK. So what are *'cells'*?

Let's go back a step - we said earlier that the mobile phone is basically a two-way radio.

The two-way radios used, for example, by the police and ambulance services, are used mainly used for contact with a central dispatcher.

Each service is allocated a small number of frequencies (or 'channels') permanently dedicated to them in each operational area.

This means that most conversations can be heard by some or even all the radios (sometimes called *'walkie-talkies'*) on a particular service's radio network.

The entire set of usable radio frequencies is often referred to as the **'frequency spectrum'**.

There is a huge demand for radio frequencies from a whole flock of users - TV and radio stations, phone companies, police, fire brigade, ambulance and the military.

Because God (in Her wisdom) only created a limited frequency spectrum, there are not enough suitable frequencies available to satisfy everybody, and so clever schemes have been developed to use the frequencies efficiently.

The mobile phone 'cell' system is one of these clever schemes.

The ingenious solution is to divide the country into a patchwork of areas called 'cells'. Each cell is the area covered by a pair of transmitting and receiving antennas.

These antennas connect your mobile to the mobile phone system. [*How these 'cells' are actually made is discussed a bit further down under 'The base station'*].

The hexagonal shape of the cells shown here is the ideal – the actual shape is a lot less regular.

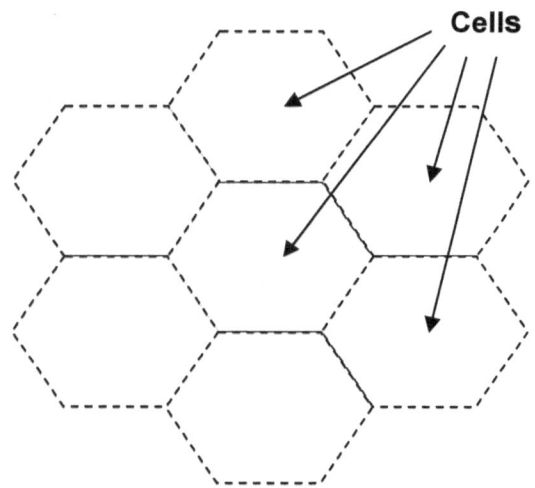

Cells

When making or receiving a call, each GSM phone within a cell is allocated a pair of frequencies, (one for transmission and one for reception) from a fixed set (or 'pool') of frequencies. At the end of the call, the allocated frequencies are returned to the pool for someone else to use.

If no spare frequencies are available in a particular area, for instance, at peak times in the centre of a city, the cells sizes are made smaller.   This is so that the same pool of frequencies is shared by a smaller number of users.

When a user in a car or bus moves from one cell into an adjacent cell, the system senses this. It then allocates a pair of frequencies in the new cell, without interrupting the call. This is helped by a small overlap between the adjacent cells.

The pool of frequencies in a particular cell is different to the pools in the adjacent cells - this is to avoid interference in the areas where the cells overlap.

However, the same pool of frequencies can be used in non-adjacent cells a bit further away.

In this way, a relatively small set of frequencies can serve a large number of customers. This is achieved:

- by using the same set of frequencies in different geographic areas; and
- by allocating channels (ie frequencies) to each phone in a particular cell **only** while a call is in progress.

A further saving of the valuable frequencies is achieved by having up to eight phones in each cell share the same frequencies – each sending and receiving their calls as short pulses in different time-slots.

The whole system of allocating frequencies and time slots is controlled automatically by very complex software and hardware - the customers are not aware of what is going on behind the scenes.

The Base Station

The most visible part of the mobile support infrastructure (and the one which causes the most aggro) is the **Base Station**.

The base station sends and receives all the radio traffic to and from the mobile phones.

It has two main parts - a set of directional antennas and a mass of control and radio electronics housed in a metal cabinet or small building nearby.

In towns and cities, the antennas are often mounted on the top of tall buildings and tend not to be noticed.

But in the suburbs and in the country, they are set on top of steel towers, mostly painted in an exciting battleship grey colour, which stick out like sore thumbs and tend to attract much criticism.

The base station electronics are generally housed in small weather-proof buildings at the base of the suburban towers.

24

Each pair of directional antennas (one to transmit and one to receive) at the top of a tower, are arranged to cover one third of a circle, so that three pairs cover the full 360 degrees.

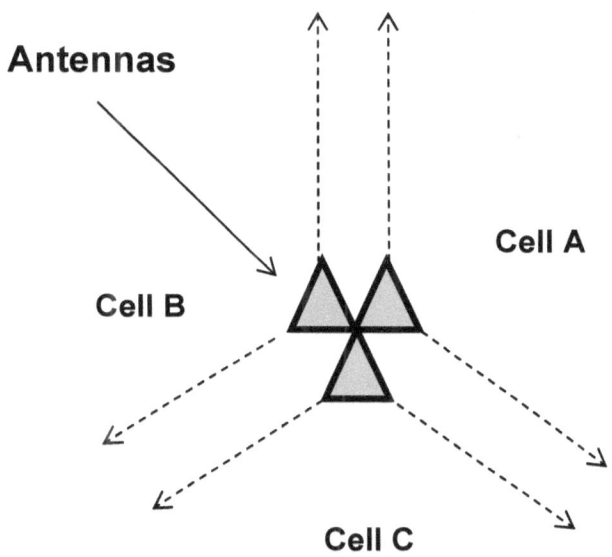

The directional antenna coverage defines the cells discussed above, so we get three cells for each base station.

With a GSM system, the cells can stretch up to **35 kilometres** from the antennas.

To make a cell smaller in a city, the antennas are tilted downwards to cover a smaller area, a bit like a flashlight beam.

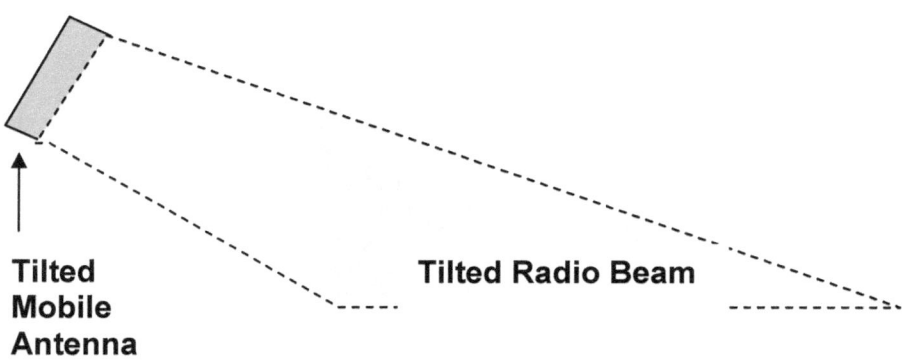

**Tilted
Mobile
Antenna**   **Tilted Radio Beam**

For special events like football finals or rock concerts, temporary base stations, mounted on trucks, are used to create extra local cells to handle the increased traffic.

Note: Be aware that our reliance on mobiles may create a problem in emergencies - particularly bushfires.

If a base station or its antenna tower is destroyed or badly damaged, mobile communications will be lost for most users being served by that tower.

Even if towers or base stations are working, severe congestion often occurs in emergencies as everyone tries to phone at the same time.

Examples of this are the **Canberra** and **Victorian** bushfires.

Base Station Controllers

The base stations are connected by fibre cables or microwave links to **Base Station Controllers** (BSC).

Depending on the system design, one controller can look after up to 512 Base Stations.

The Base Station Controllers allocate the radio frequencies at the Base Stations. They also orchestrate the exchange of voice and data between the base stations and the mobile telephone exchanges.

Base Stations

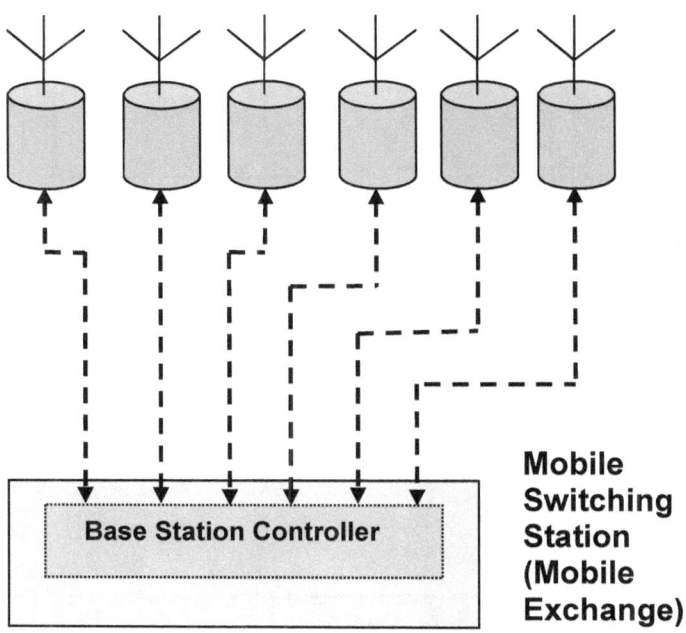

Mobile Switching Station (Mobile Exchange)

Base Station Controller

In most cases the Base Station Controllers are actually located in the mobile exchanges.

In the industry jargon, telephone exchanges are called 'switches' and mobile exchanges are known as **Mobile Switching Centres** (MSC).

The Mobile Switching Centres are where the mobile phone calls are connected to the rest of the telecommunications network (and other mobile users).

Keeping Track of Customers

Mobile phone systems have an elaborate set-up to keep track of their subscribers and make sure they are legitimate paying customers.

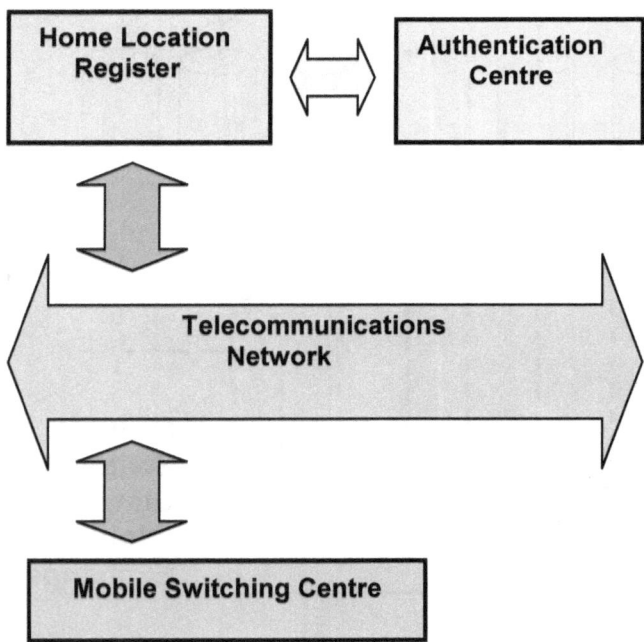

Each mobile phone company has a list of its subscribers stored on a computer system called a **Home Location Register** (HLR).

When it needs to check a subscriber, a Mobile Switching Centre accesses the HLR through the telecommunications network.

Connected to the HLR is an **Authentication Centre** (AC), which is the heart of the fraud prevention system.

Before being allowed to connect to a mobile phone company's network, the mobile must first pass a stringent test set by the Authentication Centre.

The reason for these tests is that the first generation of mobiles (like the Telstra AMPS system) had very little fraud resistance.

Hackers and criminals copied mobile phone identities with little difficulty, and ran up huge bills at other people's expense!

The same tests also allow stolen mobiles to be blocked – as many of us know.

Using your Mobile

What Happens When you Turn on Your Mobile?

When you turn on your mobile, the phone tries to establish contact with your operator (eg Telstra, Optus, Vodafone, etc.).

Your phone scans the mobile frequency band for a particular identifying signal called a **beacon frequency,** which each operator transmits in each cell.

If it cannot find your operator's beacon frequency, the display will indicate something like '*No Signal*'. This may happen in country areas or in radio 'shadow' areas caused by hills or by tall buildings in cities.

The phone companies are continually filling in poor signal areas in cities and along main highways by installing additional base stations.

If your phone finds the right beacon frequency, the display comes up with the network name, the cell name (in some cases) and a received signal strength indication (the little bars on the side of the display).

Your network then starts an authentication procedure to see if you are a valid user (and that you have paid your bills).

If you are a valid user, your voice scrambling (encryption for privacy) is set up and you are then ready to send or receive calls.

What Happens When you Make a Call?

When you want to make a call, your phone asks the system for a pair of channels from the pool of frequencies available in your cell. If a pair is available, the phone is instructed to switch to that pair.

The numbers you dial are sent through the base station to the nearest Mobile Switching Centre, (the mobile phone exchange) where the connection to the other phone is set up.

When the connection is established, your voice (converted into digital form, compressed to save bandwidth and encrypted for privacy) is sent in short pulses to the base station.

At the base station, your voice is converted to look like any normal phone conversation (in digital form). It is then transmitted to the Mobile Switching Centre, where it is switched through the network to the phone you are calling.

Ring Tones

An industry has grown up providing special ring-tones for phones – a way of individualising a mass-produced product.

Some ring-tones are unbelievably annoying - the choice of ring-tone probably says a lot about the phone owner and some have obviously not been taking their medication

Texting

One of the most popular mobile services is 'texting', a very cheap and convenient way of exchanging messages with friends or family.

Some people have calluses on their fingers from incessant texting. The strange thing is – the mobile phone system designers had no idea how popular this feature would become.

The technical term for the texting service is the **Short Message Service** (SMS).

Why is SMS so cheap? The main reason is that a text message needs only a tiny fraction of the frequency bandwidth used by voice. The SMS messages are sent on the phone's signalling channel when there is a free time-slot – SMS does not have a high priority.

With some overseas phone services you can get confirmation of the receipt of a text message, but in most cases in Australia, you send the message and hope for the best.

Phone Cameras

A feature of almost all mobiles is a still or video camera. Cameras are great for photographing or filming cats or friends.

Even though the quality is sometimes poor, phone cameras often snap pictures of fleeting incidents – TV news producers love them.

Mobile Phone Barcodes

You may have seen odd-looking patterns like this (on the right) on advertisements and business letters (as stamps). What are they and what can I do with them?

These are 'two-dimensional barcodes', which usually contain a web address or a phone number, but can contain any information, including images or music clips.

Mobile phones with a camera and special software programs can read these patterns. Many new phones have this software already installed, and you can download the special software for some of the older phones.

You can also download software to your PC to generate your own mobile codes - say your phone number or website address for your business card or flyer.

Some technology prophets think that mobile codes are going to transform the way we do things – a bit like the way texting took off.

There is more on this at the end [see **Extra Bits – Mobile Barcodes**].

Which Mobile Service Will Give me the Best Coverage?

The first thing you need to decide is where you are mainly going to use your phone - the cities or towns where you will usually be and your usual road routes. Then go to the websites of the mobile service providers and look for coverage maps of the areas you are interested in.

A list of mobile phone companies is available at http://toolkit.aca.gov.au/mobile/csp.htm. or at http://www.amta.org.au/

Phone coverage is usually shown on a colour-coded map (see below), often for both handheld phones and car-kits.

For example, the Telstra maps are at http://www.telstra.com.au/mobile/networks/coverage/maps .cfm and the Optus maps are at http://www2.optus.com.au/.

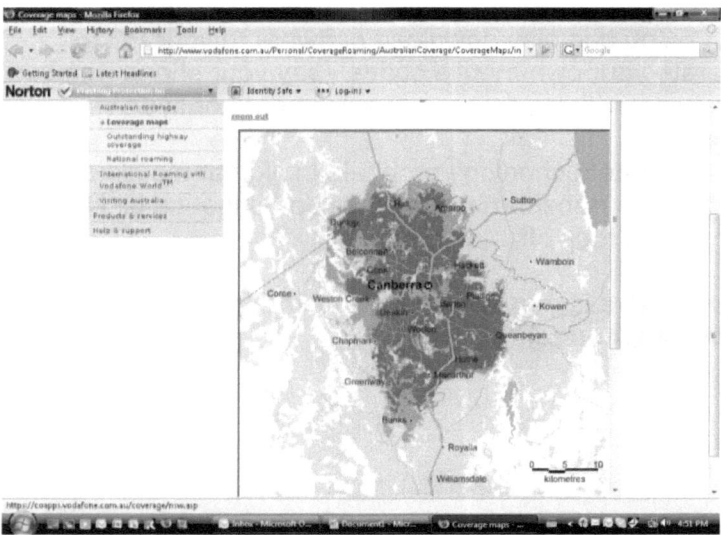

Although coverage is improving all the time, be aware that some of the coverage claims may be a trifle optimistic.

Why do the Calls Cost so Much?

The basic reason is that the carriers have spent hundreds of millions of dollars each setting up and maintaining their infrastructure, and want a return on their investment.

This infrastructure includes the exchanges, the antenna towers, the links between and the operating software.

Mobile calls are also timed - unlike local calls on a land-line in Australia. The fact that there is intense competition between mobile providers does tend to keep a lid on charges.

Service providers generally offer incentives for you to use their networks outside the peak business hours and so spread the load and reduce their infrastructure costs (which are driven by peak-time demand in the big cities).

The carriers and their retailers generally subsidise the purchase cost of the phones provided you sign with them for a minimum period (mostly a year or two).

In Australia, subsidised phones are generally 'locked', so that they only work on the supplier's network until the end of the contract[8].

There is a very brief summary of the Australian mobile phone market at the end [see **Extra Bits - The Australian Mobile Market**].

---

[8] Mobile phone 'locking' is illegal in some countries.

Faster Mobile Systems

Inevitably, many consumers wanted more than the basic GSM mobile can provide, so faster systems were developed – the so-called '**Third Generation**[9]' or '**3G**' systems[10].

These systems can handle much higher data transmission rates – theoretically almost 1000 times greater, so that Internet and video can be delivered to and from your handset (or your laptop).

The 3G phones get their higher data speed, by using a much faster **'air interface"** - the two-way radio link between the mobile phone and the base station.

In a typical 3G system, all the phones share the same frequency band. This is instead of each phone being allocated a specific pair of radio frequencies (to transmit and receive).

The phone calls are distinguished from each other by a unique transmission code - more on this below.

---

[9] GSM is considered a '**Second Generation**' (2G) system.

[10] Some 3G systems are rather grandly described as **Universal Mobile Telecommunications Systems** (UMTS).

One of the most popular 3G radio transmission systems is called **wideband code division multiple access** (W-CDMA).

There are reported[11] to be more than **7 million** 3G-capable mobile phones in Australia.

Note that the download speeds claimed by phone companies often refer to the **peak** speeds, meaning the system is having a good day and everything is working perfectly, with the wind behind it.

In practice the speeds are often a lot less – sometimes down to about half the peak.

CDMA

The original Telstra **code-division multiple access** (CDMA) system had about the same slow speed as GSM.

CDMA was used in country areas, because a small number of transceiver towers could service a very large area.   This contrasts with GSM, which is limited to a radius of about 35 kilometres from each tower.

The old CDMA system was shut down in 2008 - to allow the valuable frequency bands to be used for other radio-based services.

---

[11] According to the '*The Australian*' newspaper, Nov 17, 2008, p35.

Smartphones

The faster data rates of the 3G systems makes '*Smartphones*' possible, loaded with lots of special features. Initially very popular with business people, they are now rapidly penetrating the mass market.

An early example is the **Blackberry**, a phone that can send and receive e-mails, and has an address book, a diary and a '*to do*' list.

You can use it as your personal digital assistant (PDA).

Newer versions can access wireless Internet connections (like Wi-Fi[12] 'hotspots' in cafes and coffee shops) for faster e-mails and web browsing.

A problem for big-fingered males (or cats) is the very small number and letter buttons on some of the Smartphones.

One approach to overcoming this problem is the Apple **iPhone 3G**, which has a touch-screen instead of a number pad. It also has a camera, satellite navigation and an iPod-like music player.

---

[12] More on Wi-Fi later.

Motion sensors in the iPhone work out which way up the phone is being held and the screen changes to suit.

Another sensor knows when the iPhone is being held to your ear (as a phone) and switches off the touch screen to prevent accidental presses

When within range, the iPhone can switch to a Wi-Fi 'hot-spot' for broadband Internet connection (for e-mail and web browsing).

Because of the huge success of the iPhone[13], the competition is responding with similar phones – examples include the Blackberry *'Storm'*, HTC *'Dream'* and the Samsung *'Omnia'*.

Some Smartphones have GPS satellite navigation - there is a summary of how this works at the end – [see **Extra Bits - GPS**].

The GPS capability opens the door to all sorts of location-based applications and services on a Smartphone – like where is your bank's nearest ATM and where are the local pubs or shoe shops[14].

---

[13] *'The Australian'* newspaper reported that 200,000 iPhone sales were expected by the end of 2008. Smartphone sales in Australia have increased from 80,000 in 2003 to an expected **3,000,000** in 2008.

[14] Using the satellite navigation capability, one application (*'Show the Loo'*) gives the location of the nearest public toilet and the directions to it.

One GPS application is **Google Latitude**, which shows where yours friends or family are on your phone's screen (if your friends and family have agreed and their mobile phones are switched on at the time)[15].

Smartphone Warning

The main thing to watch with 3G Smartphones is your billing arrangement. Without some sort of cap, you can run up huge bills downloading stuff from websites.

Smartphone Crime

Press reports suggest that there is a huge market in fake Smartphones. The Australian Customs Service confiscated **3200** suspected fake mobiles in 2008.

One newspaper[16] reported that **fake** brand iPhones are supplied in China at **$18.50** each. They are then sold wholesale in Australia at **$55** each, and sold on in pubs and backpacker hostels for up to **$400** each (much less than a **real** iPhone).

A source interviewed by the newspaper claimed that up to 10,000 units were shipped at a time.

---

[15] See the article on privacy concerns in '*Geoslavery*', Dobson and Fisher, IEEE Technology and Society.

[16] '*The Australian*', 4-AUG-09, p 25.

Computer and Memory Chips

Transistors were invented in 1948 by three American physicists (**Bardeen, Brattain and Shockley**), working at the Bell Telephone Laboratories in New Jersey, USA.

Transistors quickly replaced electronic valves, and cut the power consumption and size of electronic devices enormously.

The next step was to combine a complete transistor circuit in a single chip. This was first achieved by **Jack Kilby**, of Texas Instruments, in 1958.

The first complete computer in a chip is believed to have been developed in 1968 in the US, for use in military aircraft, but was secret at the time.

Current large memory and computer chips contain hundreds of thousands of components in a finger-nail sized unit.

With impressive processing power, very small size and low power consumption, these chips have made mobile phones possible.

Special Mobiles

Rugged Phones

If you work on building sites or make a habit of getting dead drunk and dropping your mobile down the toilet, you may want to consider getting a rugged mobile phone!

These phones, which are usually coated with rubber or heavy-duty plastic, can't be used under water.

But they will survive the occasional drop into muddy puddles or being stood on. The downside is the extra size and cost.

A basic rugged GSM phone, like the **Samsung M110**, costs about $200.

A rugged 3G phone, like the **Sony Ericsson C702** (picture on the left), will cost a bit more - about $600.

A rugged Smartphone, like the **Motorola NC75**, which is basically a hand-held computer (with built-in GPS capability) combined with a phone, costs a lot more – around $3,200.

Satellite Phones

If you regularly travel to remote areas outside the range of your normal mobile, you might consider a satellite phone ('*satphone*').

Most satphones are bigger and heavier than ordinary mobiles and generally have a big antenna sticking out the top or the side.

The radio signals to and from satphones, are passed through dedicated satellites circling the earth.

Very few places on the planet are outside the range of these satellites.

The downside of satphones, (apart from the size and weight) is the cost. New phones cost almost $2,000 each, and call plans range from $125 to $185 per month.

The good news is a thriving market for second-hand satphones, which are a lot cheaper than new ones.

Satphone services available in Australia include **Iridium**, **Globalstar**, **Inmarsat** and **Thuraya**.

Some possible future satphone developments are discussed in **Part 4** below.

The Dangers of Using a Satphone
(if you're a rebel leader)

In November 199, after the
Soviet Union collapsed,
Dzhokhar Dudayev (right,
in the big hat), became
President of the
breakaway Chechen
Republic .

According to press reports, Dzhokhar was vapourised on
April 21, 1996, by two laser-guided missiles when his
location was detected by Russian reconnaissance planes[17]
while using a satellite phone.

---

[17] Reportedly Sukhoi **Su-24MR** and **Su-25**.

Mobile Phones in the Developing World

According to the International Telecommunications Union (ITU), there were **3.3 billion** mobiles in use in the world at the end of 2007 – about 1 phone for every two people.

In the two years to the end of 2007, India had added 154 million phones, while China had added 143 million.

The biggest impact was in Africa, with an annual growth rate of **49%**. Mobiles account for nearly **90%** of African phone subscribers.

The advantage of mobile systems in developing countries, is that they are much cheaper and quicker to roll out.

This is because they don't need thousands of kilometres of wire or optical fibre to make it happen.

Mobile Phones and Politics

Capacitors are components which can store electrical energy for short periods. They are used in many electronic products (including mobile phones), for filtering signals.

Capacitors using the element **tantalum,** can be made very small. They are very often used in things like phones and iPods. The rising demand for miniature electronics has driven up the price of tantalum.

While tantalum is produced in several countries, Australia produces most of the world's supply - mined by Talison Minerals[18], and The Sons of Gwalia[19].

Only about 1% of the world's supply of tantalum comes from the Congo. There it is extracted from a dull black ore called columbite-tantalite, usually referred to by the local term '**coltan**'.

However, some lobby groups and journalists claim that most of the fighting in the Democratic Republic of Congo is driven by a desire to profit from that country's mineral riches. These riches include gold, diamonds and tantalum.

It is claimed that the Rwandan army and some rebel groups have made fortunes illegally exporting tantalum from the Congo.

[18] The mine is at Wodgina, about 100 kilometres south of Port Hedland in Western Australia.

[19] Now in receivership. The mine is located about 5 kilometres west of Leonora in Western Australia.

Environmental groups are also worried about the impact of illegal coltan mining.

They are particularly worried about the negative effect on the **East Lowland Gorilla**[20] (on the right) - a threatened species.

Electronics manufacturers have been asked by lobby groups not to buy tantalum sourced from the Congo – a very hopeful ask!

---

[20] What is believed to be an East Lowland Gorilla called '*Conor*' has been seen using a mobile phone in Melbourne.

PART 2 - WILL MOBILE PHONES HARM ME?

Mobile phones are an indispensible part of a huge number of people's work and social life, but could there be a downside? For some years, concerns have been raised that using mobile phones can cause brain tumours.

At regular intervals, TV current affairs programs such as '60 Minutes' and 'Lateline' raise the health issues. For example, on 'Today Tonight' on June the 12th, 2008, David Smith expressed his anger at the lack of research into the health effects of mobile phones.

When he was 30 years old, David had a tumour 'as big as a golf ball' removed from behind his right ear around the acoustic nerve. He had been a mobile phone salesman for 10 years and had used mobiles from 1 to 2 hours each day.

On the same program, a leading Canberra neurologist, Dr. Richard Bit-Tar, appeared. He has written a research paper on the link between mobiles and brain tumours.

He recommended that the Australian Government and the phone industry take immediate steps to reduce the exposure of consumers to mobile phones. While he says that it is hard to point the finger solely at mobiles, he recommends using mobiles sparingly.

In reply, a spokesman for the Australian Mobile Telecommunications Association (AMTA), Chris Althus, said that he was not worried about getting a tumour.

So what can we make of all this? Lets look at what researchers and regulators think so far.....

Heating effects

Regulatory bodies setting exposure limits for mobile phones, focus on the **heating effects** of the body by the radio signals emitted by the phones. These limits are generally given as the maximum permitted power, absorbed per unit mass of our bodies.

The measure widely used is the 'whole body **specific absorption rate**' (SAR). In Australia, this is currently 1.6 Watts per kilogram[21]. The UK limit is 0.4 Watts per kilogram.

The International Commission for Non-Ionising Radiation (ICNIRP) recommends an SAR of 0.06 Watts per kilogram for the whole body and 2 Watts per kilogram averaged over any 10 grams of tissue.

---

[21] The Cancer Council notes that a draft Australian Standard is being developed and that the recommended limit (SAR) may be raised to 2 Watts/kilogram.

Mobile phones vary widely in the level of SAR they produce, but the newer phones are thought to be generally better than the old. There is no independent measurement of SAR under standard conditions.

A test commissioned by the BBC-TV '*Panorama*'[22] program in 1999 showed big differences between phones, as shown in the table below:

Phone	SAR (Watts/Kg)
Nokia 2110	0.44
Nokia 5110	0.37
Nokia 6110	0.29
Bosch World 718	0.28
Ericsson GA628	0.26
Hagenuk Global Handy	0.03
Motorola V3688	0.02
Motorola Star Tac 70	0.02

An extensive list of SAR measurements for most popular phones is provided by the Mobile Manufacturers Forum at http://www.mmfai.org/public/sar.cfm/lang=eng.

The Australian Communications and Media Authority (ACMA) provides health information on its website at http://www.acma.gov.au/WEB/STANDARD/pc=PC_2052

---

[22] http://news.bbc.co.uk/1/hi/health/351048.stm

The Australian Government body which looks at radiation is the Australian Radiation Protection and Nuclear Safety Agency (ARPANSA). It notes that:

'A few animal studies suggest that exposure to weak microwave fields can accelerate the development of cancer.'

But overall, the ARPANSA conclusion is that:

*'There is no clear evidence in the existing scientific literature that use of mobile telephones poses a long-term public health hazard (although the possibility of **a small risk cannot be ruled out**).'*

The Cancer Council of NSW (http://www.cancercouncil.com.au) has a '***Position Statement***' on mobiles which discusses health concerns.

The Council's basic position is similar to ARPANSA - that at the frequencies and at power levels permitted for mobiles in Australia, there is no evidence **yet** of harmful effects.

But, the Council is also cautious:

'The lack of evidence does not.... prove the absence of a risk and more specific research is needed.' and that *'...as relatively little is known on the long term effects of ...exposure, more research is needed in this area'*.

The Cancer Council also suggests that:

'......some people may wish to minimise their exposure to [radiation] when they use a mobile phone.'

The World Health Organisation (WHO) likewise recommends caution, suggesting that:

'...concerned people should limit their own and their children's...exposure by limiting the length of calls and using a "hands free" device to keep the mobile phone away from their head and body.'

Despite the enthusiasm of the WHO, tests by the Consumers Association in the UK suggest that, while some **'hands free'** kits can cut down the radiation exposure, other kits can increase it. Their effectiveness depends on many factors.

Given that their power levels are lower than the mobiles, one would expect the heating effect of devices like *'Bluetooth'* headsets to be negligible.

Base stations

With the Base Stations, it is only the transmitting antennae that are of concern. Given that they are located on tall towers or tall buildings, the Cancer Council notes that the base station radiation:

'.....at ground level and in regions normally accessible to the public are many times below hazard levels and no heating effect has been detected'.

Other biological effects

Some researchers suggest that radio waves can affect our cells in ways other than through heating.

Dariusz Leszczynski, at the Radiation and Nuclear Safety Authority in Finland, exposed 10 female volunteers to radio waves equivalent of an hour-long GSM phone conversation.

He compared 580 different proteins in the volunteer's skin cells, and found that one protein has increased by **89%** and another had decreased by **32%**[23].

Although similar effects have been observed in cultured cells in laboratories, this is first time such effects have been reported in living humans.

Australian Research

The Australian Government (through the National Health and Medical Research Council) is funding research[24] through a centre of excellence - the **Australian Centre for Radiofrequency Bioeffects Research** (ACRBR).

The Centre is a consortium of Monash University[25], RMIT University, Swinburne University of Technology and Telstra Research Laboratories in Melbourne, and the Institute of Medical and Veterinary Sciences in Adelaide.

---

[23] Reported in the *'New Scientist'* magazine, 23-FEB-08.

[24] With an allocation of **$2.5 million**, over five years.

[25] Through its Department of Epidemiology and Preventative Medicine.

Since Telstra Research Laboratories are providing much of the technical know-how, critics are concerned that the involvement of a major mobile phone provider, may compromise the independence of the Centre's research.

The Centre's pronouncements to date have been fairly cautious, and echo the 'wait and see' approach of the WHO and other Australian Government agencies.

Large Scale Studies

The largest epidemiological study conducted so far is the INTERPHONE study[26], which covered 13 countries, including Australia, over 12 years.

The Australian Cancer Council has considered the INTERPHONE report and issued some summary observations on 10-MAY-2010. The Cancer Council said:

- *[The INTERPHONE study]* found no evidence that **normal use** of mobile phones, for a period up to 12 years, can cause brain cancer.

- However, the study found that in a small subset of patients with glioma[27], their tumour was more likely to be on the same side of the head as the mobile phone was used - where there was **excessive use** (**30 minutes a day or more**).

---

[26] See the United Nations **International Agency for Research of Cancer** (IARC) website: www.iarc.fr/en/research-groups/RAD/current-topics.php.

[27] Glioma is a cancer of the glial cells. The insulating myelin sheath, which covers the axons (the connections in the brain between the neurons), is made of glial cells.

- While this doesn't prove a link between brain cancer and mobile phones, it does point to a need for more investigation of heavy phone use.

- The study involves phone usage for 12 years at most, so it tells us little about risk associated with mobile phone use over decades.

- In particular, insufficient time has passed since mobile phones were introduced, to determine whether or not there is a risk to children. This INTERPHONE study will continue to assess use beyond the current 12-year time span.

- Until this area has been fully investigated, Cancer Council recommends that if parents remain concerned about possible effects on children - **they should minimise the children's use of mobile phones and encourage hands-free/speaker options, or texting.**

- Anyone concerned about the harmful effects of electromagnetic energy should reduce their use of mobile phones, or employ hands-free technology.

What If?

Overall, there appears to be indications that **heavy** mobile phone use may present some health risks. If a convincing case is established for this, the major hurdles for health and regulatory authorities will be:

- our reluctance to reduce our dependence on a really useful technology;

- the inevitable rear-guard action conducted by a huge industry and its vested interests; and

- our mostly subjective perception of risk. We rarely take action based objectively on statistics. We all tend to discount risks associated with the things we are familiar with – generally along the lines of *'It won't happen to me'*.

However, if a potential health problem is identified, it may prompt phone manufacturers to design phones with safer antennas or better hands-free gear.

What should we do?

So what can we conclude from all this?

Most authoritative Australian and international bodies say there is no convincing evidence yet that there is a problem for most of us. But they cannot rule out a risk for heavy users. They all urge minimum use of mobiles as a precaution.

At this stage, the sensible and safest things to do would seem to be:

- **Use your mobile as little as possible for making and answering voice calls;**

- **Keep your voice calls (making or answering) as short as possible;**

- **Delay giving your kids a mobile for as long as possible. When they get one, restrict their use to texting and emergency calls only. This is if you can – good luck with that one!**

But wait………….. there's more (which may be very important for your health)…….

Mobile Phones and Road Accidents

MOBILE PHONE HEALTH WARNING

One way in which mobile phones **can** damage your health is in **road accidents.** Many reputable authorities warn about **the increased risk of accidents** for drivers using mobile phones, even with 'hands free' kits – read on.......

Authorities have been warning drivers for some time.

In August 1999, the Royal Society for the Prevention of Accidents (RoSPA) in the UK, pointed to tests carried out in a driving simulator.

The tests, by the Psychology Department of Aston University, showed that dangers existed, even with 'hands free' phones. Remember this was in 1999!

An RoSPA Road Safety Adviser, Dave Rogers, said:

'The tests we used in the [mobile phone] conversations, showed that using a hands-free phone was no less distracting than using a hand-held phone.

There is no basis to claim one system is safer than the other.

*It is the **distraction of the conversation** rather than the mechanics of using the phone that poses the real problem.'*

In 2003, the mobile phone use of **17,023 drivers** in metropolitan Melbourne, was observed in research by Associate Professor David Taylor, (Director of Emergency Medicine at the Royal Melbourne Hospital) and colleagues[28].

The researchers found that 18.5 drivers per thousand (**1.85 per cent**) used hand held mobile phones while driving.

Men had a slightly higher rate of use than women, while older drivers had a significantly lower rate than middle-aged or young drivers.

They found that the rate of mobile phone use was significantly higher in the evening compared with the morning.

According to the researchers, there was increasing evidence that the use of a hand-held mobile telephone, while driving a vehicle, increases the risk of an accident.

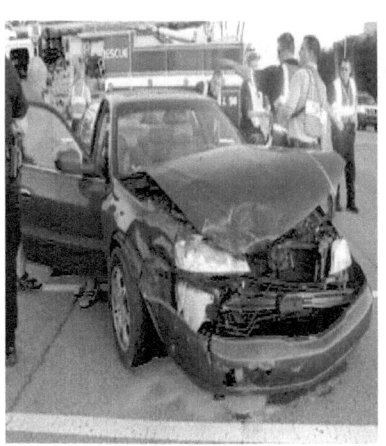

They claimed that the risk of collision while using a hand-held mobile phone increased four-fold.

---

[28] Published in the *'Medical Journal of Australia'* in August 2003.

This compares to a two-fold increased risk when driving with an illegal **blood alcohol concentration of 0.06 per cent**.

A case-control study showed that **the risk of a fatality is increased nine-fold when using a hand-held mobile phone.**

Professor Taylor said:

'This is likely to represent a preventable cause of injury, as mobile phone use causes driver inattention and increases accident rates,'

Anne Bolling, a researcher at the Swedish National Road and Transport Research Institute, reported in January 2004, that people get distracted when they're speaking on the phone. But they also compensate by slowing down and driving more carefully.

The study looked at a small sample of 48 drivers (half using hand-held and half using hands-free phones). Each person drove in a simulator for about an hour, and received 10 phone calls that lasted about a minute each.

Bolling insisted that hands-free devices were just as likely to distract a driver's concentration:

'It is distracting to speak on the phone, but it makes no difference if you're using a regular [hand-held mobile] phone or a hands-free.'

In fact, her research showed that drivers trying to dial a number on a hands-free phone were more likely to be distracted than those using a hand-held phone.

'It's worse dialling on a hands-free, because you have to look down beside the steering-wheel,' she said.

One could expect similar findings for texting. Research data on texting while driving is hard to find. But based on Anne Bollings' research, one could speculate that it is at least as dangerous as talking on a mobile.

In July 2005, the British Medical Journal published a study of more than 450 drivers in Western Australia.

These drivers owned or used mobile phones and had been involved in car crashes serious enough to warrant hospital treatment.

Most of the drivers also gave their permission to get records of the phone use from their mobile phone company. This was to get precise timings of calls and see how these compared to the estimated time of the crash.

The study found that drivers who had used a mobile phone - either holding it to their ear or using a hands-free system - were **4.1 times more likely** to have an accident in the next 10 minutes than if they had not made a call.

The risk with a hand-held phone was **4.9 times**, and with hands-free phones it was **3.8 times** – so hand-free phone seem to be better, but not much.

Using a hand-held mobile phone while driving, is illegal in all Australian states, but every day we see people driving while talking or texting on their hand-held mobiles, and prosecutions seem to be relatively rare!

Although there are no studies available at the time of writing, anecdotal evidence suggests that GPS navigation units in cars are also very distracting, and may have been the primary cause of some fatal accidents.

The evidence to date suggests that we could have a lot more to fear from mobile-distracted drivers than from a mobile-induced cancer.

How should we respond to all this?

If you are driving, before setting out on your journey, set up:

- your radio;

- your CD player;

- your GPS navigator (with voice prompt activated); and, most importantly

- **set your mobile to voicemail.**

PART 3 – THE FUTURE – FASTER AND HIGHER

'The future is already here – its just not evenly distributed'

William Ford Gibson (The science-fiction writer who coined the term *'çyberspace'*)

3G Broadband

Wi-Fi

Wi-Fi (pronounced 'Y Fye') is the brand name of an international standard for wireless broadband access[29].

The standard is based on an invention by Australian scientists and engineers from the **CSIRO**[30] Radiophysics Division, and derived from CSIRO's work in radio astronomy.

By early 2009, the patents (after lengthy and expensive court cases against 14 multinational technology companies) had yielded **$200 million** to CSIRO, with a lot more to come.

---

[29] The standard is excitingly known as **IEEE 802.11**. The name 'Wi-Fi' was devised by the brand consulting firm, **Interbrand**, alluding to the term Hi-Fi in audio equipment.

[30] The (Australian) Commonwealth Scientific and Industrial Research Organisation

Most 3G Smartphones and laptops have a **Wi-Fi** capability built in. This gives your Smartphone or laptop broadband Internet access, when you are within range of a Wi-Fi '*hotspot*' – a wireless network access point. 'Hotspots' are usually at airports or near cafes or coffee shops.

Probably more than 95% of the Australian population now has broadband coverage for 3G mobile phones (and laptops), using enhancements like **EDGE** (Enhanced Data for GSM Evolution) or **HSPA** (High Speed Packet Access).

HSPA can down-link (to your phone or laptop) at up to **14.4 Megabits per second**[31], and up-link (from your phone or laptop) at up to **5.76 Megabits per second**. This is for fast Internet browsing, or video download (like YouTube).

Development is under way for even faster speeds. The next version of HSPA, called **Evolved HSPA** or **HSPA+,** aims to provide up to **22 Megabits per second** up-link and up to **42 Megabits per second** down-link[32].

---

[31] Under ideal conditions, with the wind behind it.

[32] Again under ideal conditions.

4G Development

In the medium to long term, competition is under way between at least two very high-speed **4G** systems:

LTE

Another 4G development under way is **LTE** (Long-Term Evolution), which aims to bring Internet protocols[33] (such as TCP/IP) right down to the mobile handset or laptop.

LTE proposes to use some very fancy (and cat impressing) technology such as:

- Orthogonal Frequency Division Multiple Access (OFDMA);

- Multiple Input – Multiple Output (MIMO);

- Spatial Division Multiple Access (SDMA); and

- antenna beam forming.

OFDMA is claimed to be very efficient in its use of the radio spectrum (delivering lots of data in a limited bandwidth).

LTE eventually aims to provide data speeds of up to 300 Megabits per second.

An advantage of LTE for GSM and 3G carriers is that quite a lot of the existing infrastructure can be used for the new system.

---

[33] **Internet protocols** are the rules that govern data transmission on the Internet.

WiMAX

One 4G system which already exists is **WiMAX** (Worldwide Interoperability for Microwave Access).

This provides up to **70 Megabits per second** for both up-link and down-link, (and may provide up to 100 Megabits per second in the future).

WiMAX can be used for mobile phones and to deliver broadband Internet where cables are not available or are too expensive to install[34].

A disadvantage of WiMAX systems, is that they are not compatible with existing cell phone infrastructure, and have to be set up from scratch.

Wateen Telecom is installing a WiMAX system in Pakistan, which will eventually cover 71 cities.

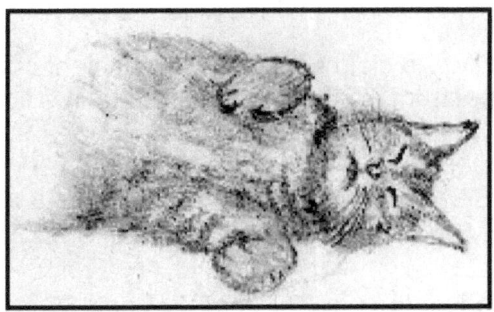

---

[34] WiMAX systems were used in Aceh, Indonesia after the tsunami and after Hurricane Katrina in the US to provide broadband Internet access for emergency services.

Some (Possible) Satellite Phone Futures

The Iridium System

The **Iridium** satellite phone system, previously mentioned, can be used in Australia,.

It was launched in 1998 by a consortium led by Motorola. The system has 66 working satellites, (like the one on the right) and 13 spares in low earth orbit.

The satellites are each zipping about 800 kilometres above your head, and moving at 27,000 kilometres per hour.

The system covers Australia (and the world), even in the most remote deserts and on offshore oil platforms. Phones and plans can be bought through Telstra.

The original consortium went broke after a year, and was put up for sale at a greatly reduced price. The system was bought by a US-led consortium, which included the Perth-based Quadrant (with a 16% share).

Quadrant is now Fulcrum Equity in Melbourne and its shareholding has been sold down to 6%.

In September 2008, Iridium announced a plan to merge with a US investment bank. It was intended to pay off Iridium's debts, and raise $US2 billion for a replacement system (to be called 'Iridium Next').

The replacement system, scheduled to be complete by 2016, aimed to have a data rate up to **10 megabits per second.**

The bad news for the Iridium system was a collision in space on 10-FEB-09, in which an Iridium 33 satellite was hit by a runaway Russian military satellite (Kosmos 2251) and totally destroyed.

The Thuraya System

In Australia, Optus markets the **Thuraya** satellite service. The Thuraya-3 satellite covers Australia and the surrounding region.

Thuraya[35], based in Abu Dhabi, uses three large Boeing communications satellites[36] in geosynchronous orbit. Geosynchronous satellites move so that they appear stationary to users on the ground. A single geosynchronous satellite can serve a large area.

The dual-band Thuraya phones (such as the SG-2520) are about the same size as normal mobile phones, and can use both the Thuraya and the Optus GSM networks. The SG-2520 costs about $1,386 new.

Optus access rates for the Thuraya service are about $50 per month and call rates are about $1.30 per minute to Australian mobiles and landlines.

---

[35] Thuraya is the Arabic term for the *Pleiades* star constellation.

[36] The Thuraya satellites were launched using a Ukrainian '*Zenit*' rocket from a modified oil-drilling platform, '*Ocean Odyssey,*' floating in the sea on the equator, near Kiribati. The launch company, a consortium of Russian, Norwegian, Ukrainian and US interests, went into US Chapter 11 bankruptcy in mid-2009.

PART 4 – MORE DETAIL

Part 1 was an overview of how a typical mobile phone system works. This section expands on Part 1, starting with why the world's most popular mobile phone system, the European GSM, was developed in the first place.

Why GSM?

In the early 1980's there were five different (and incompatible) mobile phone systems in Europe. This lack of harmony disturbed some members of the evolving European Community.

A committee was set up (the **Groupe Special Mobile** or **GSM**) to oversee the design of a new European mobile phone system[37]. The committee gradually expanded until it included engineers, standards experts, regulators, phone companies and manufacturers.

Australia was the first non-European member of the group.

After nearly 10 years of effort, the first GSM systems started operating. After the usual teething troubles, the system quickly matured, and systems are now available in all parts of the world.

---

[37] Set up under the **Conférence Européenne des Administrations des Postes et des Télécommunications** (CEPT) in 1982. In 1987, 13 countries agreed to develop a European cellular system. In 1989, responsibility for the GSM project was transferred to the **European Telecommunications Standards Institute** (ETSI). The first part of the GSM standard was issued in 1990.

For marketing purposes, GSM now stands for **Global System for Mobiles**.

The first GSM operator was **Radiolinja** in Finland in 1991, using a Nokia-supplied network.

The first commercial GSM call was made by the (then) Finnish Prime Minister, **Harri Holkeri**, using a Nokia mobile.

Holkeri didn't just sit around making phone calls. He chaired the United Nations General Assembly in 2000-2001.

As a member of the negotiating team led by former US Senator George Mitchell, he played a vital role in arranging the Good Friday Agreement in British-occupied Ireland.

He was awarded an honorary knighthood by Queen Elizabeth II of England.

What did the GSM Design Team Aim for?

The GSM group set out to design a system, which would work across the whole of Europe, and overcome the shortcomings of the existing systems.

The main problems were:

- fraud (criminals copying mobile phone numbers and billing calls to unsuspecting owners);

- short battery life;

- inefficient use of the crowded radio spectrum and

- privacy (people could easily listen to conversations with cheap 'scanners' - as politicians, royalty and celebrities found out).

The main design aims for the GSM system were:

- ability to use the same handset in any country;

- resistance to fraud;

- better privacy for users

- efficient use of the radio spectrum; and

- longer battery life.

To encourage competition between operators, the system was designed to allow up to **five** operators in any area.

To encourage manufacturing competition, it was decided that the GSM infrastructure would be '**modular**' – that is, a complete system could be made up from parts bought from different manufacturers.

To achieve this, the inputs and outputs of the various component 'boxes' were carefully defined, but **how** the insides of the 'boxes' were designed or worked was up to the individual manufacturers.

The next sections look at what goes on inside a typical GSM handset.

Warning

Readers allergic to acronyms or diagrams should seek medical advice before proceeding beyond this point.

The GSM Handset

The part of the mobile you can see in your hand – the handset, with its keypad and screen, has its own unique user identity – excitingly known as the **International Mobile Equipment Identity (IMEI)**. It is usually found printed on a label underneath the battery.

If your phone is lost or stolen, you can report the loss to your local police by quoting the IMEI number, which is generally also found on the side of the box in which your phone came.

Useful Hint

Keep the box in which your mobile came, just in case your phone is lost or stolen. The empty box is also a great place to keep cat treats.

Voice processing

A normal land-line telephone, carries voice frequencies from 300 Hertz (cycles per second – more about this later) to 3,400 Hertz – this is referred to as the 'audio bandwidth'.

Although it sounds 'tinny', this audio bandwidth does allow voice recognition. GSM mobiles use the same audio bandwidth as land-lines - **3,100 Hertz**.

When we use a digital mobile, our voice is converted by a small microphone, from a pressure (sound) wave, into an electrical wave.

In conventional digital sampling, this electrical wave is 'sampled,' (like the way a slightly open Venetian blind 'samples' the view through a window) and the samples are then converted into digital form – with numbers representing the height of the wave at the sample instant.

Voice Wave

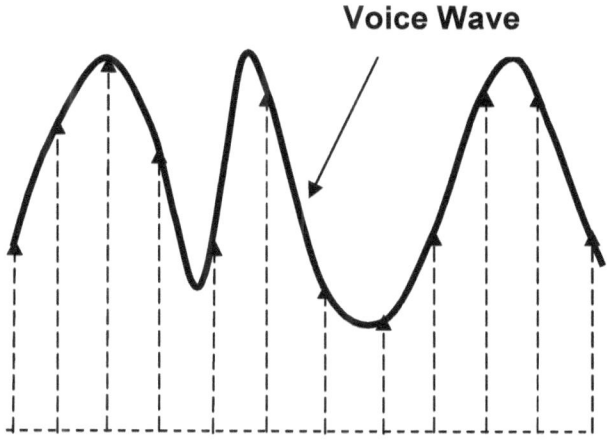

Sample Intervals

The problem with the conventional approach, is that to get acceptable voice quality, the sample rate has to be, at the very least, **twice** the highest voice frequency, and preferably much more than that - this is in order to be able to follow the ups and downs of the voice wave.

Thus, for a 3,100 Hertz audio bandwidth, (the normal telephone voice bandwidth), the samples would have to be at least **6,200 samples per second**.

The amount of digital data generated by this approach becomes far too large for an efficient mobile phone, so we need a better method.

GSM phones use an amazing technique called **Linear Predictive Coding** (LPC), which, in effect, creates a digital 'model' of our vocal system.

In the phone, the electrical wave corresponding to your voice (or a cat meow) is sampled at only **50 samples per second**.

Each digital sample is fed into a special LPC processor (a small computer), which analyses the voice sample for three characteristics:

- **loudness**;

- **pitch** (voice frequency); and

- **'formants'** (the individual distinguishing characteristics of each voice).

Using the loudness, pitch and formants data - an approximation of our speech is calculated, and compared with the actual speech sample. The difference between the two is called the **'residue'**.

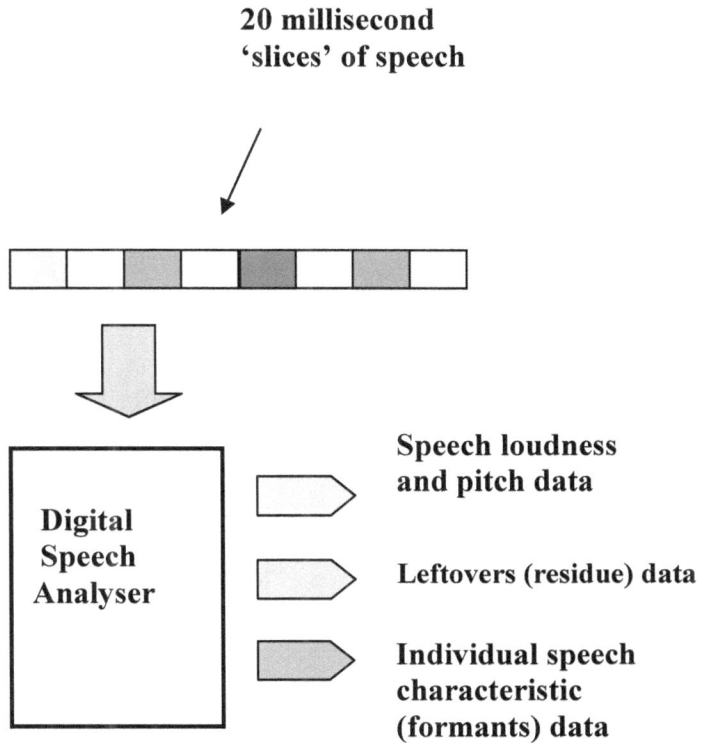

20 millisecond 'slices' of speech

Digital Speech Analyser

Speech loudness and pitch data

Leftovers (residue) data

Individual speech characteristic (formants) data

SPEECH ENCODING

The loudness, pitch, formant and residue data for the sample are stored in memory, and transmitted by the radio section of the phone in short pulses at a rate of 217 pulses per second.

At the receiving phone, the whole process is reversed.

In the LPC processor of the receiving phone, the formants data is used to create a special filter, which mimics the characteristics of the original speaker's voice.

The loudness, pitch and residue data are fed into this filter and a re-created voice sample emerges. This process is very efficient and produces a recognisable version of the original voice.

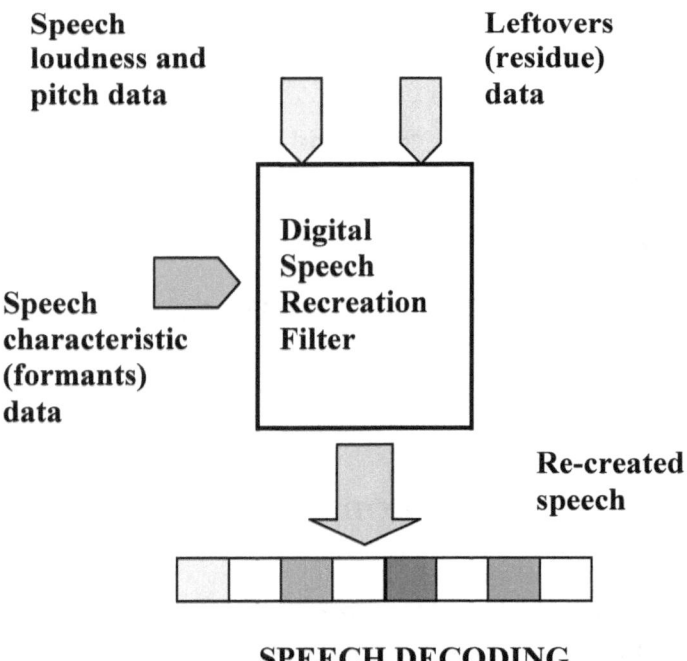

SPEECH DECODING

To save the battery, the mobile transmits only when there is something to transmit.

When there is no incoming voice, during a mobile call, so-called 'comfort noise' is inserted (a sort of very low 'swoosh'). This is because the dead silence was found to be very disconcerting for users.

If a transmitted pulse is not received for some reason, in most cases the hearer will not notice.

This is because some of the samples are spread across a number of pulses, and a sophisticated error-correction scheme allows the missing samples to be restored.

The Radio Section

This section looks at how your mobile transmits and receives voice, pictures, video and data.

Background on radio waves

Electro-magnetic radiation includes radio waves, light and x-rays.

Lower frequency radio waves (like those used for AM radio stations) can 'bend' around buildings and hills to some degree.

At higher frequency (such as those used for TV, FM radio and mobile phones), they tend to behave more like light beams, and cannot 'bend' around obstacles (although they can reflect off buildings).

Radio frequencies used

Two radio frequency bands have been allocated to GSM in most parts of the world (including Australia) – one around **900 MegaHertz** and the other around **1800 MegaHertz**.

For the basic Australian GSM systems, the **transmitting** frequencies (i.e. from the phone), are from **935 to 960 MegaHertz**, while the **receiving** frequencies (i.e. to the phone), are from **890 to 915 MegaHertz**.

Phone companies are normally allocated **5 MegaHertz** in each band, so up to five carriers can operate in any area at the same time.

The 5 MegaHertz transmitting and receiving bands for each carrier allow 124 phone channels (paits of frequencies), each spaced 200 kiloHertz[38] apart.

Each channel has 8 time slots, allowing 8 phones to operate at the same time on the same channel.

Remember that the channels can be used over and over again in different cells, provided adjacent cells do not use the same channels (because they would interfere where the cells overlapped).

Radio Communication

Radio communications were pioneered by **Guglielmo Marconi**, with his radio telegraph system.

He shared a Nobel Prize for physics in 1909 (even though he did not seem to have made any original contribution to science).

Marconi had an Italian father and an Irish mother (the granddaughter of the founder of the Jamison's Irish whiskey business) and was brought up in Italy.

A committed fascist, Marconi was a Senator in the Italian parliament and a member of the Italian Fascist Grand Council.

The Italian dictator Benito Mussolini was best man at Marconi's second marriage in 1927.

---

[38] A kilohertz is a thousand Hertz (cycles per second).

Hertz

MegaHertz means millions of cycles per second.

The term Hertz is used in honour of the German discoverer of radio waves, **Heinrich Hertz** (1857 - 1894) - the attractive gentleman on the left.

Heinrich, who was professor of physics at Karlsruhe Polytechnic in Germany, died of blood poisoning at the age of 36.

Heinrich's nephew **Gustav Hertz** won a Nobel Prize for Physics in 1925 – obviously a smart family.

Gustav Hertz's son, **Carl**, was a soldier in the German army in WWII. While he had the bad luck to be on the losing side, he did have the good fortune to be captured by the US Army (rather than the Soviet Army) and was shipped overseas.

A friend of his father pulled strings and arranged for him to go to Sweden, where he became a professor at the University of Lund.

While at Lund, Carl invented the medical ultrasound machine.

Modulation

The radio section of the mobile phones sends and receives all information (voice, pictures or data), in digital form. The information is encoded on the radio signal using **frequency modulation**[39] (i.e. the radio frequency is varied to replicate the information).

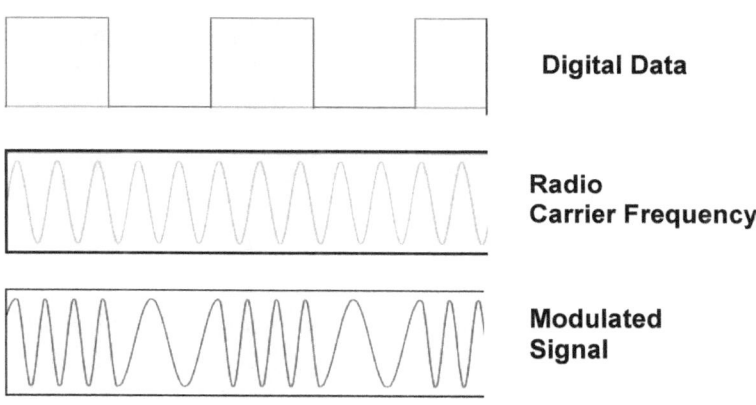

Digital Data

Radio
Carrier Frequency

Modulated
Signal

FREQUENCY MODULATION

At the receiving end, the radio signal is demodulated - that is, the process is reversed and the digital data is extracted from the radio signal.

---

[39] The modulation scheme used is **Gaussian Minimum Shift Keying** (GMSK).

Pulsing

The radio transmission and receiving signals are in the form of short pulses at 217 pulses per second. Pulsing has two advantages.

The first advantage is frequency sharing - by timing the pulses correctly, eight mobiles can use the same frequency. This process is known as **time division multiple access** (TDMA).

The second advantage is battery saving. Because the mobile is transmitting only one eighth of the time, it is **not transmitting** seven-eights of the time. The downside is the increased complexity of the system and some interference with other electronics – like the buzzing noise you sometimes hear in your radio.

You may note signs in hospitals and doctors waiting rooms, asking you to turn off your mobile. This is because the transmitted pulses of your phone can interfere with sensitive medical equipment, like heart monitors and ECGs.

Transmission power

The transmitting power of a mobile handset (which is an important factor in the health debate) is limited to a maximum of **2 Watts** in the 900 MegaHertz band and **1 Watt** in the 1800 MegaHertz band.

Watts

Watts are named in honour of **James Watt** (1736 – 1819), a Scottish mechanical engineer and inventor.

James didn't invent the steam engine, but greatly improved its efficiency.

His improvements made a major contribution to the Industrial Revolution.

Frequency hopping

Sometimes radio signals bounce off buildings and hills in such a way that they come together, and cancel each other out at a particular spot.

To counter this problem, the mobiles can change frequency in successive transmissions (and receptions), a procedure called *'frequency hopping'*.

If a pulse is not received using a particular frequency, the next pulse, at a different frequency, has a better chance of being received, and the 'lost' pulse can be recovered using error correction routines.

The 'Hyperframe'

All this frequency sharing and hopping, means that mobiles can transmit and receive in a large number of combinations of both time and frequency.

Making sure that no two mobiles try to use the same frequency, at the same time, in the same cell, is a complex task. It is one of the reasons why GSM system software is very expensive.

The starting point is a big table (called a '**hyperframe**'). This is a list of all the possible combinations of time and frequency which can be allocated to the mobiles.

The system rolls through the entire hyperframe in just over two hours and then repeats itself.

SIM Card

The part which gives the handset its individual personality, is the **Subscriber Identity Module** (SIM).

This is a tiny 'smart card' with your subscriber details; special phone numbers (for instance for emergencies); secret numbers (for instance, 'keys' for scrambling your voice) and memory space for your phone book and text messages.

How some of these features are used is described a little later.

The removable SIM card is placed in a special socket inside the handset.

In GSM-speak, your handset, complete with a SIM card, is known as a **Mobile Station** (MS).

Data Connections

For mobile office work, there is a small socket on the bottom or side of (most) handsets for a data link to a laptop computer.

The socket connects by cable to a credit-card sized card (known as a PCMCIA II card), which plugs into the side of your laptop.

This arrangement allows the laptop to send and receive data calls.

These days, most of the more expensive laptops have a wireless connection already installed.

Newer mobiles have a short-range, low-power radio link (for example, *'Bluetooth'*) to link to a headset without using a cable.

SMS

A very important feature of all mobile systems is the **Short Message Service** (SMS), which allows text messages (up to 132 characters long) to be sent or received.

The SMS data is transmitted at relatively slow rates using the signalling channel - when a signalling time slot is available. This is instead of using the more valuable voice channels, and is why SMS messages are so much cheaper to send.

Texting is creating new (abbreviated) versions of all languages, to the dismay of the 'educated'!

When You Switch On

The beacon frequency

To allow the mobiles to connect to their carrier's system, each phone company transmits a special signal in each cell, called a 'beacon frequency'.

When first switched on, the mobile searches for the right beacon frequency, which contains all the information the mobile needs to get started. Like all the other frequencies, the beacon frequency has eight time slots.

The first time slot (t_0)[40] is a reference signal, which synchronises the mobile's frequency, with the master frequency in the base station.

The mobile is then told the current state of the 'hyperframe', i.e. where the system is in its two-hour cycle of time and frequency slots.

The mobile continuously monitors the strength of the beacon frequency signal. It uses this information to adjust the mobile's transmit power up and down - to give the best performance (and save the battery).

The mobile is also monitoring the received level of beacon frequencies of the nearby cells, so that it can switch to another cell - in, say, a moving car - when the adjacent cell level is better. The 'handover' from one cell to another is so fast[41] that the user is normally completely unaware that it has occurred.

---

[40] In techno-speak, this is pronounced 'T Zero'.

[41] GSM lore suggests that the very fast hand-over was at the insistence of the French delegation, which wanted no discernible loss of signal as the French TGV high-speed trains passed from one cell to the next.

When a phone changes cells, the cell location number (LAI) is stored in the SIM card, and also sent back to the mobile exchange, so the system knows where you and your phone are.

WARNING

The following section is complicated. Patients should be seated comfortably in quiet surroundings. Alcohol intake should be strictly limited to aid concentration.

Patients with short attention spans should consider taking a nap before starting.

The next sections look at:

- How the system checks for valid phones (bills paid, not lost or stolen); and

- How your voice privacy is set up.

Authentication

Authentication is the main anti-fraud defence of the mobile system. The authentication has **two stages** – a check of a valid phone list and a check of the phone itself. It is complicated but fast, so the user is only aware of a slight delay.

To start the authentication, the mobile is instructed to transmit the unique **International Mobile Station Identifier** (IMSI) stored on its SIM card - incidentally this is the only time that the IMSI is transmitted in unencrypted (i.e. not scrambled) form.

The receiving base transceiver station (BTS) sends the IMSI through the host **Mobile Switching Centre** (MSC) to the **Home Location Register** (HLR) - which may be in another city.

The HLR is a large database, which holds all the phone company's subscriber details.

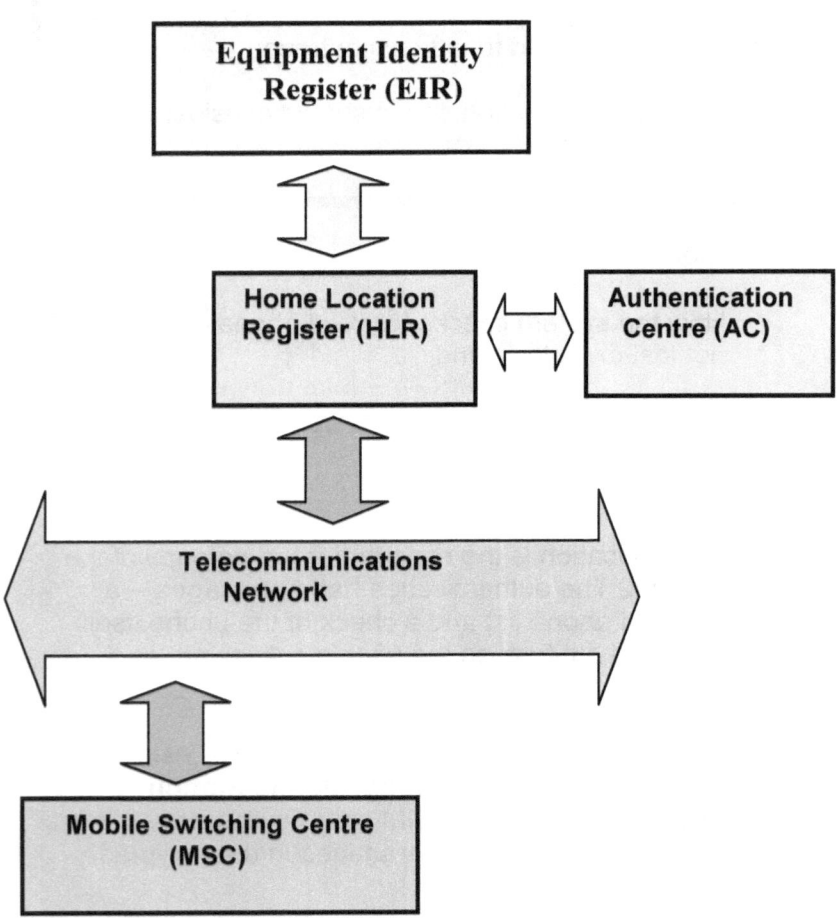

In the Home Location Register (HLR), the first part of the authentication uses the phone IMSI to check for the user's rights, (and to see if the phone is stolen), in the **Equipment Identity Register** (EIR).

If all is well, the **Authentication Centre** (AUC), attached to the HLR, starts the second phase of the authentication, by doing three things:

- It generates a 128-bit random number - '**RAND**';

- It retrieves a special 64-bit number called a 'key' **(Ki)** - unique to the requesting mobile; and

- It generates a set of dummy identifying numbers (known as **Temporary Mobile Station Identifiers** or TMSI) for the mobile. The dummy numbers are used instead of the mobile's real identity (IMSI), in all subsequent transactions, to increase security.

RAND, Ki and the set of dummy numbers are sent back to the requesting Mobile Switching Centre (MSC), where they are stored in the **Visitor Location Register** (VLR).

The VLR is a local data register within the MSC, which contains information relating to every mobile currently communicating with that MSC. This information includes the cell in which each phone is located.

The second part of the phone authentication starts when the RAND number is transmitted to the mobile.

Both the mobile and the VLR then perform the same calculation, using RAND and Ki (which the mobile already has stored in its SIM card), in a special procedure ('*algorithm*') excitingly called '**A3**'.

Ki
(unique mobile key number)

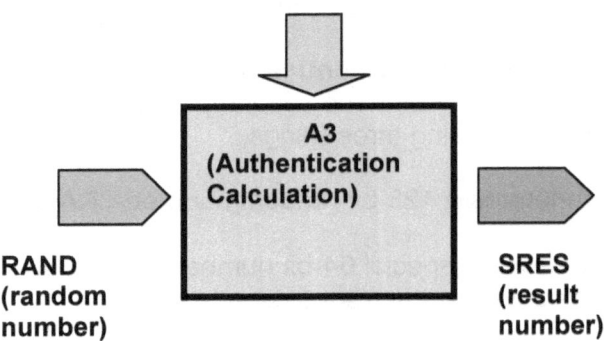

If the result (called **SRES**) is the same for both the phone and VLR calculations, the mobile is considered authentic.

Voice security

The voice security is provided to stop others listening in to your conversations using 'scanners'.

Firstly, both the mobile and the VLR calculate a 64-bit number called a 'cipher key' (**Kc**) using the RAND number in a procedure (algorithm) called '**A8**'.

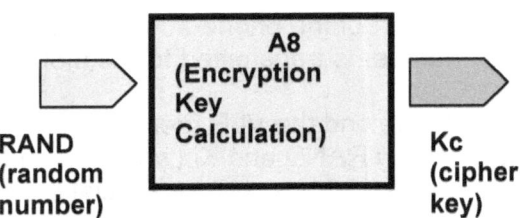

The VLR sends the cipher key (Kc) to the **Base Transceiving Station** (BTS), which commands the mobile to start transmitting in cypher (ie encrypted) mode.

The mobile and the BTS then both use Kc as the key to encrypt all voice and data traffic, using an encryption procedure (algorithm) called '**A5**'[42].

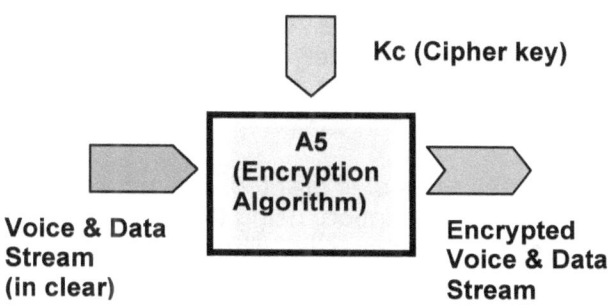

Note that only the 'air path' (ie the radio link between the mobile and the BTS) is encrypted.

---

[42] There are several variants of the A5 algorithm - A5/0 provides no encryption; A5/1 is the original algorithm used in Europe; A5/2 provides weaker encryption for export and is used in the US; A5/3 provides the strongest encryption and is used in Australia.

When You Make a Call

When you make a call, the mobile uses the signalling channels for call set-up.

The mobile first tries to send a request on the **random access channel** (RACH) time slot on the beacon frequency. If another mobile is already sending a request, the mobile waits until the RACH is free.

When the request is received by the system, the mobile is instructed to immediately switch to the **access grant channel** (AGCH) time slot, to keep the RACH free for other requests.

If the access request is granted, the mobile will be sent its frequency pair allocation, using the **dedicated control channel** (DCCH) time slot. The mobile immediately switches to this frequency pair (send and receive) and the call commences.

When Someone Calls You

When someone calls you on your mobile - say from a conventional phone - the call is passed to your phone company's Home Location Register (HLR).

The HLR knows which mobile exchange (MSC) you were last connected to. The call is switched to that MSC, and the Visitor Location Register (VLR) checked to see which **cell** your mobile is in.

A 'paging' call, [using the **paging channel** (PCH) time slot on the beacon frequency], is sent by the Base Transceiver Station (BTS), which covers your cell.

If your mobile is switched on and you are not already making a call, the call is switched through to you.

Otherwise the caller is sent a busy tone; a 'not available' message (not switched on or out of range of the system) or the caller is diverted to a message bank.

International Calls

'International Roaming'

If you make a call to an overseas visitor using his own mobile in Australia, the call is first sent to a local Australian Home Location Register (HLR).

The HLR will recognise the visitor's country of origin and carrier from the number.

The call set-up is diverted through an international gateway to the visitor's home HLR. The visitor's home HLR will have a record of the last mobile exchange (MSC) used by the visitor in Australia.

The system tries the indicated MSC in Australia, and if the visitor's mobile is still in that area, the call is set-up in the usual way.

If you are overseas, using your Australian mobile, the same sort of thing happens. If your carrier has a 'roaming agreement' with a phone company in the country in which you are travelling, your mobile will connect to that local phone company's network.

When you first switch on in the country you are visiting, the host GSM network in that country will recognise your origins. The network will call your home HLR in Australia for authentication.

When verified, your home HLR will send the RAND number and dummy identifiers to your current host MSC. In the MSC, they are stored in the Visitor Location Register (VLR). From there on, the procedure is the same as when you were at home.

For details of international coverage and partner services, try:

www.gsmworld.com/roaming/gsminfo/index.shtml

The Billing System

A large and vital part of all mobile phone systems, is the billing software, (particularly from the phone company's point of view)!

The billing system must know when your calls occur, how long they last and the tariff applying at that time.

It must collate all this at regular intervals, and send out (hopefully accurate) bills to you, or subtract the correct amount from your pre-paid account.

Phone Bill Alert

Changing over from one billing system to another almost always results in major billing hiccups for the mobile phone companies – **always** (but always) check your bill thoroughly after a billing system change or upgrade.

It's not that the phone companies are trying to cheat you – it's usually because all the bugs have not been ironed out of the new billing software.

More on 3G Systems

As we said previously, GSM allocates a pair of frequencies (transmit and receive) to each phone in a cell, with up to eight phones sharing the same frequency pair, in a scheme called **time division multiple access** (TDMA).

The problem for the original GSM system, is that it has a very limited bandwidth, and cannot handle reasonable speed Internet or video. A more effective method to achieve high data speeds, used on many 3G systems, is **code division multiple access** (CDMA).

CDMA is a '**spread spectrum**' system. This means that instead of each phone transmitting and receiving on specific allocated frequencies, the signal burst from each phone is spread out across a range of frequencies.

This signal burst is spread across the full bandwidth allocated to the phone service provider, usually 5 MegaHertz for the transmit channel, and 5 MegaHertz for the receive channel.

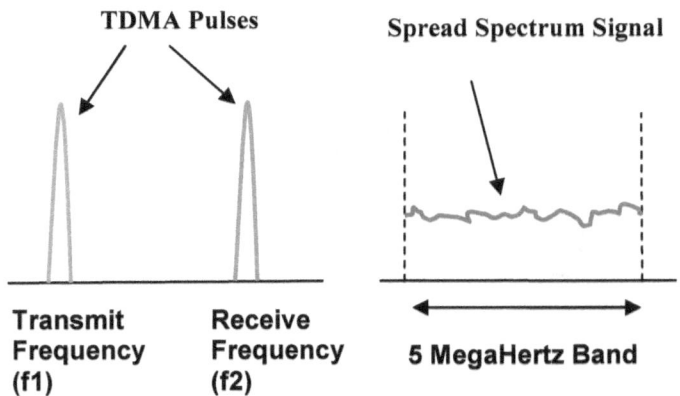

With CDMA - a unique transmission code is generated for each phone. This code is in the form of a random sequence of zeroes [0] and ones [1], called a **pseudo-random direct sequence**. The sequence is generated at a rate[43] much faster than the data[44] to be transmitted.

The data (voice, pictures or Internet) is then mixed with the random sequence, and transmitted as a short burst. The transmitted signal looks very like static or 'white noise'.

At the receive end, the signal burst is fed into a small computer. This computer rejects all signals - except the one which exactly matches the transmission code. The processor then extracts the data from the selected signal.

An analogy sometimes used to explain CDMA - is our ability to pick out and lock on to a conversation, in a language we understand - in a room full of people speaking different languages.

The particular form of CDMA used by most current 3G systems is called '**Wideband CDMA**' (WCDMA or W-CDMA).

Apart from the data speed, the radio part is the principal difference between GSM and 3G. The rest of the underlying infrastructure is similar.

---

[43] The rate is called the 'chip rate' and each code bit (0 or 1) is called a 'chip'.

[44] Data in used here in the sense of **any** digital data – it could be digitised text, web pages, voice, video or still pictures.

FAMOUS SPREAD SPECTRUM INVENTORS

An early US patent relating to spread spectrum was granted on 11-AUG-1941 to the Hollywood film actress **Hedy Lamarr** (1913 – 2000) - picture at right - and her friend, the pianist and composer **George Antheil** (1900 – 1959).

Hedy Lamarr was born Hedwig Kiesler in Vienna. From 1933 to 1937, she was married to an Austrian (fascist) arms manufacturer, **Freidrich Mandl**.

Mandl disapproved of her acting in risque films, and tried to stop her by taking her to meetings with technicians and businessmen. It was at some of these meetings that she learned about *'frequency hopping'*, which is related to spread spectrum.

The daughter of Jewish parents, Hedy fled Austria with all her jewellery (but minus the husband) in 1937, first to Paris and then the US.

Mandl left Austria for Brazil and then Argentina, where he served for a time as an advisor to the dictator, Juan Peron (Eva's husband). He later returned to Austria.

Congestion

A feature of CDMA, is the way congestion - too much phone traffic - is handled.

In a GSM system, when all the frequencies and time slots in a cell are occupied, new users are locked out and will not be able to make calls. They will be politely asked, by an automated voice, to try again later.

In a CDMA or W-CDMA system, as the cell becomes congested, the noise level increases. Users encountering this will have to make their own decision – put up with the noise or try again later.

In Australia, three 3G systems are operating – **Telstra**, **Optus** and the **Vodafone/Hutchinson** joint venture, **VHA**.

The frequency bands used[45] for 3G are different to those used by GSM.

---

[45] The Telstra 3G network uses the **850 MegaHertz** and **2100 Megahertz** frequency bands. The Optus 3G service uses only the **2100 MegaHertz** band.

Initially, these 3G networks had a download (to the phone) data speed of up to **1.5 Megabits** (millions of data 'bits') per second and an upload (from the phone) speed of up to **384 kilobits** (thousands of data 'bits') per second. A data 'bit' is a 0 or a 1.

The Telstra 850 MegaHertz '*NextG*' network increased the download data speed to **14.4 Megabits per second**, with future upgrades aiming for **40 Megabits per second**. Optus plans to match this speed from 2009.

99

EXTRA BITS

To impress your cat even more!

Why Lithium-Ion Batteries Run Out of Puff

Atoms have a positively charged and relatively heavy middle bit (the **nucleus**). The nucleus is surrounded by a cloud of negatively charged, and much lighter, **electrons**.

The negative charge of the electrons, balances the positive charge of the nucleus, making the whole atom electrically neutral.

If an atom loses electrons for any reason, it becomes positively charged and is called an '**ion**'.

Charging

In a Lithium-Ion battery, the positive terminal (the '**cathode**') is usually made of layers of lithium cobalt dioxide.

When the battery is being charged, the lithium atoms in the cathode are stripped of their electrons. The resulting **positively charged** lithium ions move inside the battery from the cathode, through a separator screen, and pile up on the **negative** battery terminal (the '**anode**').

At the anode, which is usually made of porous graphite (a form of carbon), the arriving lithium ions find stray electrons hanging about, and become complete lithium atoms again.

Discharging

When the battery is delivering power to a mobile phone, the lithium atoms piled up at the anode - lose their **negatively charged** electrons into the electronic circuits, providing the electric current.

The resulting **positively charged** lithium ions then leave the anode, and drift through the inside of the battery back to the cathode, where they are re-absorbed.

Time and tide wait for no battery

You can expect a Lithium-Ion battery to withstand up to **5,000** or so charge cycles - which should last you the life of your phone.

However, with constant charging and discharging, the gaining and losing of lithium at the cathode, causes the lithium cobalt dioxide layers to degrade. This leads to a build up of impurities in the cathode, so that it loses its capacity to re-absorb the lithium. If this happens, it is time to get a new battery.

Mobile Barcodes

We are all familiar with **barcodes**, like the one on the right. These so-called '*one dimensional*' (1D) barcodes contain information about an item, the manufacturer and the country of origin.

When the checkout in a shop scans the barcode, it not only brings up the price, but in most cases also updates the stock re-ordering system.

3G mobiles with a camera and suitable software, can read '*two-dimensional*' (2D) barcodes (often just referred to as '***mobile codes'***), to read in phone numbers, text, website addresses or whatever.

Many new phones have the barcode reading software already installed[46], and you can download suitable software for some older phones.

There are two common standards for 2D barcodes:

---

[46] Telstra has been installing free 2D code reading software in all its 3G phones since June 2008.

Data Matrix Code

     The Data Matrix code, developed by Acuity CiMatrix[47], conforms to international standard **ISO/IEC16022**, and is in the public domain - meaning that it can be used free of any licensing or royalties.

     The Data Matrix symbols (like the one on the right), are recognizable by the solid L-shaped border on the left and bottom. You may see these on business letters as stamps.

Use a reader on your camera phone to decode me...

     Each Data Matrix symbol, can encode up to **3,116 alphanumeric characters** (numbers, letters, punctuation and symbols), and is surrounded on all sides by a blank border called a *'quiet zone'*.

     The newest version of Data Matrix (ECC200) supports error checking, which allows the recognition of barcodes that are up to **60% damaged**.

---

[47] Formerly a division of **Robotic Visions Systems** of Nashua, New Hampshire, USA and owned by **Siemens** of Germany since 2005.

Quick Response (QR) Code

The QR (Quick Response) code was originally developed by **Denso Wave**[48], as a high-speed code for managing car parts.

The code is in the public domain (no royalties or licensing required).

The QR symbols (like the one on the right), which are easily recognizable by the three 'squares within squares' at the corners, can encode alphanumeric characters, music, images, web addresses and emails.

Conforming to an international standard (**ISO/IEC 18004:2006**), QR is used on most Japanese-made cell phones[49].

The QR symbols can encode up to **4,296 alphanumeric characters.**

With the highest level of error correction (Level 'H' Reed-Solomon) up to **30%** of damaged code words can be recovered.

Although most are square, the Data Matrix and QR symbols don't have to be like that - some are rectangular.

---

[48] A subsidiary of **Toyota**.

[49] And is the standard supplied with Telstra 3G phones.

The Australian Mobile Market

The Australian mobile market is estimated to be worth about **$13 billion** a year, or about **1.2%** of GDP and growing.

The principal players are **Telstra**, **Optus** and **VHA**.

The estimated market shares are as follows:

Company	Customers	Revenue	Profit (EBITDA)
Telstra	42.2%	43.3%	52.5%
Optus	32.7%	29.4%	32..5%
VHA	25.2%	27.3%	25.4%

Source: 'The Australian'', 10-FEB-09

Telstra is a publicly owned Australian company, in the top 10 by market capitalisation on the Australian Stock Exchange.

Optus is wholly owned by Singapore Telecom (SingTel), the largest company by market capitalisation on the Singapore Stock Exchange.

VHA was formed in February 2009 with the merger of Vodafone and Hutchinson operations in Australia.

- **Vodafone** is owned by Vodafone PLC of the UK, which claims to be the world's largest mobile phone company.

- **Hutchinson** is owned by Hutchinson Whampoa of Hong Kong, a diverse company (with interests including ports, property, hotels, retail, telecommunications, electricity, water and transportation) which claims an annual turnover of $US 40 billion.

Note: A large number of smaller mobile phone players act as resellers, buying air-time from the main operators.

GPS

The **Global Positioning System (GPS)** is a navigation system, which uses radio signals from satellites to calculate your position anywhere on the Earth

The system was set up by the United States Department of Defence, and is managed by the US Air Force 50th Space Wing.

The reason GPS can be incorporated in mobile phones, is that the GPS receiver-processors electronics have been shrunk, and are now a tiny 15 X 17 millimetres, and will probably get smaller.

How GPS Works

The GPS system uses a set of **24 satellites** orbiting the earth[50]. Each satellite transmits radio signals, which allows GPS receivers to determine their location and speed, plus the exact time.

A GPS receiver works out its position, by processing the signals received from at least **four** of the GPS satellites. The satellite orbits are arranged, so that at least four satellites, are always visible from any point on the earth.

Each satellite carries an extremely accurate clock, and continually transmits messages containing the **exact time** and the satellite's **exact position**.

---

[50] Now increased to 31 to improve coverage and reliability.

For each satellite the GPS receiver can 'see', the receiver measures the **time** taken for messages to travel from the satellite.

From this time, the receiver calculates the **distance** to each satellite. This is possible because the speed of a radio signal - the same as the speed of light - is well known[51].

Using the distances from the known positions of the four satellites, the receiver calculates its location.

The receiver position is displayed, usually on a moving map. Many GPS units also show direction and speed - calculated from position changes.

Note that the GPS phone **must** have a clear view of the sky – it will not work in buildings or trams.

Once your position is known by your GPS Smartphone, all sorts of other applications – like where the nearest shoe shop or pub is – allows you to swing into action.

Positioning

You might think that three satellites are enough to give your position, since space has three dimensions.

However, very small errors in the receiver clock, multiplied by the very fast speed at which satellite radio signals move, can result in large positional errors.

The receiver uses a fourth satellite to correct the receiver clock errors, and give a more accurate position.

---

[51] In space, the speed of light is **299,792.458** kilometres per second. In air it is slightly slower, depending on the density.

CONCLUSION

How technology works may be seem complicated, but it is not beyond anybody's understanding. It is **definitely not** just the preserve of a select group.

Our understanding of technology, and how it might affect us, is important for us to make informed decisions about how and when we want to use it.

While the current evidence suggests that the health dangers for average mobile users is low, all authorities suggest caution in using mobiles until we know more – particularly for the young. We don't really know how mobile use will affect the health of the young three decades from now.

But the current evidence strongly suggests that mobiles are dangerous when we drive - even with 'hands-free' systems. We could expect that **all distractions** – like CD players and GPS units – will present dangers.

If this small book persuades even one person to change his or her ways, this author will be very satisfied (and so will his cat).

FURTHER READING

- The complete set of GSM technical manuals (recommended for people with no social life);

- *'The GSM System for Mobile Communications'*, Michel Mouly and Marie-Bernadette Pautet, &Sys, [At $US299, this one is not cheap!];

- *''Introduction to Mobile Telephone Systems: 1G, 2G, 2.5G and 3G Technologies and Services''*, (e-book), Lawrence Hart and David Bowler, Althos Publishing, 2008.

SOME ACRONYMS

3G	Third Generation
AC	Authentication Centre
AGCH	Access Grant Channel
AMTA	Australian Mobile Telecommunications Association
ARPANSA	Australian Radiation Protection and Nuclear Safety Agency
AUC	Authentication Centre
BSC	Base Station Controller
BTS	Base Transceiver Station
CDMA	Code Division Multiple Access
CEPT	Conférence Européenne des Administrations des Postes et des Télécommunications
DCCH	Dedicated Control Channel
EIR	Equipment Identity Register
ETSI	European Telecommunication Standards Institute
FM	Frequency Modulation
GMSK	Gaussian Minimum Shift Keying
GPS	Global Positioning System
GSM	Global System for Mobiles
HLR	Home Location Register
HSDPA Access	High-Speed Downlink Packet
ICNIRP	International Commission for Non-Ionising Radiation
IMEI	International Mobile Equipment Identity
IMSI	International Mobile Station Identifier

ITU	International Telecommunications Union
LAI	Cell Location Number
LCD	Liquid Crystal Display
Li-Ion	Lithium Ion
LPC	Linear Predictive Coding
MS	Mobile Station
MSC	Mobile Switching Centres
Ni-Cad	Nickel-Cadmium
Ni-MH	Nickel Metal Hydride
PCH	Paging Channel
RACH	Random Access Channel
RoSPA	Royal Society for the Prevention of Accidents
SAR	Specific Absorption Rate
SIM	Subscriber Identity Module
SMS	Short Message Service
TDMA	Time Division Multiple Access
TMSI	Temporary Mobile Station Identifiers
UMTS	Universal Mobile Telecommunications System
VLR	Visitor Location Register
WHO	World Health Organisation
W-CDMA	Wideband Code Division Multiple Access

www.ingramcontent.com/pod-product-compliance
Lightning Source LLC
Chambersburg PA
CBHW022026170526
45157CB00003B/1369